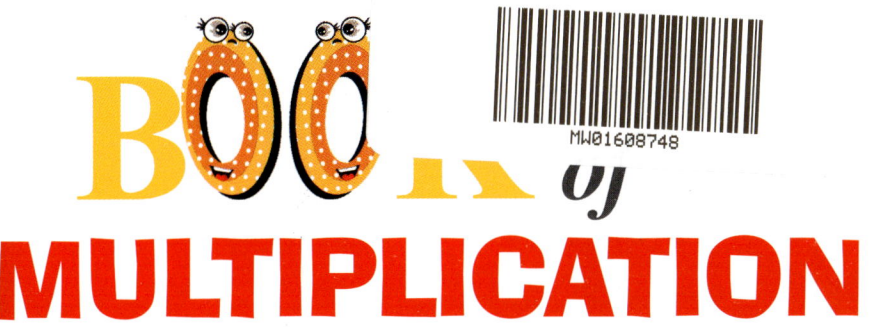

BOOK of MULTIPLICATION

2 ÷

1

3 × 4

MAPLE KIDS

Contents

Multiplication Table   3

Multiplication Tables 1 to 40   4-28

Number System   29-30

Divisibility Rules   31-32

Book of Multiplication

Published by

MAPLE PRESS PRIVATE LIMITED

Corporate & Editorial Office
A 63, Sector 58, Noida 201 301, U.P., India

phone +91 120 455 3581, 455 3583
email info@maplepress.co.in
website www.maplepress.co.in

Reprint 2020

ISBN: 978-93-50336-32-8

Printed in Noida, India

10 9 8 7 6 5 4 3

MULTIPLICATION TABLE

	1	2	3	4	5	6	7	8	9	10	11	12	13	14	15	16	17	18	19	20
1	1																			
2	2	4																		
3	3	6	9																	
4	4	8	12	16																
5	5	10	15	20	25															
6	6	12	18	24	30	36														
7	7	14	21	28	35	42	49													
8	8	16	24	32	40	48	56	64												
9	9	18	27	36	45	54	63	72	81											
10	10	20	30	40	50	60	70	80	90	100										
11	11	22	33	44	55	66	77	88	99	110	121									
12	12	24	36	48	60	72	84	96	108	120	132	144								
13	13	26	39	52	65	78	91	104	117	130	143	156	169							
14	14	28	42	56	70	84	98	112	126	140	154	168	182	196						
15	15	30	45	60	75	90	105	120	135	150	165	180	195	210	225					
16	16	32	48	64	80	96	112	128	144	160	176	192	208	224	240	256				
17	17	34	51	68	85	102	119	136	153	170	187	204	221	238	255	272	289			
18	18	36	54	72	90	108	126	144	162	180	198	216	234	252	270	288	306	324		
19	19	38	57	76	95	114	133	152	171	190	209	228	247	266	285	304	323	342	361	
20	20	40	60	80	100	120	140	160	180	200	220	240	260	280	300	320	340	360	380	400

$1 \times 1 = 1$	1
$1 \times 2 = 2$	$1 + 1 = 2$
$1 \times 3 = 3$	$1 + 1 + 1 = 3$
$1 \times 4 = 4$	$1 + 1 + 1 + 1 = 4$
$1 \times 5 = 5$	$1 + 1 + 1 + 1 + 1 = 5$
$1 \times 6 = 6$	$1 + 1 + 1 + 1 + 1 + 1 = 6$
$1 \times 7 = 7$	$1 + 1 + 1 + 1 + 1 + 1 + 1 = 7$
$1 \times 8 = 8$	$1 + 1 + 1 + 1 + 1 + 1 + 1 + 1 = 8$
$1 \times 9 = 9$	$1 + 1 + 1 + 1 + 1 + 1 + 1 + 1 + 1 = 9$
$1 \times 10 = 10$	$1 + 1 + 1 + 1 + 1 + 1 + 1 + 1 + 1 + 1 = 10$

MULTIPLICATION TABLE OF 2

2 x 1 = 2		2
2 x 2 = 4		2 + 2 = 4
2 x 3 = 6		2 + 2 + 2 = 6
2 x 4 = 8		2 + 2 + 2 + 2 = 8
2 x 5 = 10		2 + 2 + 2 + 2 + 2 = 10
2 x 6 = 12		2 + 2 + 2 + 2 + 2 + 2 = 12
2 x 7 = 14		2 + 2 + 2 + 2 + 2 + 2 + 2 = 14
2 x 8 = 16		2 + 2 + 2 + 2 + 2 + 2 + 2 + 2 = 16
2 x 9 = 18		2 + 2 + 2 + 2 + 2 + 2 + 2 + 2 + 2 = 18
2 x 10 = 20		2 + 2 + 2 + 2 + 2 + 2 + 2 + 2 + 2 + 2 = 20

MULTIPLICATION TABLE OF 3

3 x 1 = 3	3
3 x 2 = 6	3 + 3 = 6
3 x 3 = 9	3 + 3 + 3 = 9
3 x 4 = 12	3 + 3 + 3 + 3 = 12
3 x 5 = 15	3 + 3 + 3 + 3 + 3 = 15
3 x 6 = 18	3 + 3 + 3 + 3 + 3 + 3 = 18
3 x 7 = 21	3 + 3 + 3 + 3 + 3 + 3 + 3 = 21
3 x 8 = 24	3 + 3 + 3 + 3 + 3 + 3 + 3 + 3 = 24
3 x 9 = 27	3 + 3 + 3 + 3 + 3 + 3 + 3 + 3 + 3 = 27
3 x 10 = 30	3 + 3 + 3 + 3 + 3 + 3 + 3 + 3 + 3 + 3 = 30

MULTIPLICATION TABLE OF 4

4 x 1 = 4	4
4 x 2 = 8	4 + 4 = 8
4 x 3 = 12	4 + 4 + 4 = 12
4 x 4 = 16	4 + 4 + 4 + 4 = 16
4 x 5 = 20	4 + 4 + 4 + 4 + 4 = 20
4 x 6 = 24	4 + 4 + 4 + 4 + 4 + 4 = 24
4 x 7 = 28	4 + 4 + 4 + 4 + 4 + 4 + 4 = 28
4 x 8 = 32	4 + 4 + 4 + 4 + 4 + 4 + 4 + 4 = 32
4 x 9 = 36	4 + 4 + 4 + 4 + 4 + 4 + 4 + 4 + 4 = 36
4 x 10 = 40	4 + 4 + 4 + 4 + 4 + 4 + 4 + 4 + 4 + 4 = 40

MULTIPLICATION TABLE OF 5

5 x 1 = 5	5
5 x 2 = 10	5 + 5 = 10
5 x 3 = 15	5 + 5 + 5 = 15
5 x 4 = 20	5 + 5 + 5 + 5 = 20
5 x 5 = 25	5 + 5 + 5 + 5 + 5 = 25
5 x 6 = 30	5 + 5 + 5 + 5 + 5 + 5 = 30
5 x 7 = 35	5 + 5 + 5 + 5 + 5 + 5 + 5 = 35
5 x 8 = 40	5 + 5 + 5 + 5 + 5 + 5 + 5 + 5 = 40
5 x 9 = 45	5 + 5 + 5 + 5 + 5 + 5 + 5 + 5 + 5 = 45
5 x 10 = 50	5 + 5 + 5 + 5 + 5 + 5 + 5 + 5 + 5 + 5 = 50

6 x 1 = 6	6	
6 x 2 = 12	6 + 6 = 12	
6 x 3 = 18	6 + 6 + 6 = 18	
6 x 4 = 24	6 + 6 + 6 + 6 = 24	
6 x 5 = 30	6 + 6 + 6 + 6 + 6 = 30	
6 x 6 = 36	6 + 6 + 6 + 6 + 6 + 6 = 36	
6 x 7 = 42	6 + 6 + 6 + 6 + 6 + 6 + 6 = 42	
6 x 8 = 48	6 + 6 + 6 + 6 + 6 + 6 + 6 + 6 = 48	
6 x 9 = 54	6 + 6 + 6 + 6 + 6 + 6 + 6 + 6 + 6 = 54	
6 x 10 = 60	6 + 6 + 6 + 6 + 6 + 6 + 6 + 6 + 6 + 6 = 60	

$7 \times 1 = 7$		7
$7 \times 2 = 14$		$7 + 7 = 14$
$7 \times 3 = 21$		$7 + 7 + 7 = 21$
$7 \times 4 = 28$		$7 + 7 + 7 + 7 = 28$
$7 \times 5 = 35$		$7 + 7 + 7 + 7 + 7 = 35$
$7 \times 6 = 42$		$7 + 7 + 7 + 7 + 7 + 7 = 42$
$7 \times 7 = 49$		$7 + 7 + 7 + 7 + 7 + 7 + 7 = 49$
$7 \times 8 = 56$		$7 + 7 + 7 + 7 + 7 + 7 + 7 + 7 = 56$
$7 \times 9 = 63$		$7 + 7 + 7 + 7 + 7 + 7 + 7 + 7 + 7 = 63$
$7 \times 10 = 70$		$7 + 7 + 7 + 7 + 7 + 7 + 7 + 7 + 7 + 7 = 70$

8 x 1 = 8	8
8 x 2 = 16	8 + 8 = 16
8 x 3 = 24	8 + 8 + 8 = 24
8 x 4 = 32	8 + 8 + 8 + 8 = 32
8 x 5 = 40	8 + 8 + 8 + 8 + 8 = 40
8 x 6 = 48	8 + 8 + 8 + 8 + 8 + 8 = 48
8 x 7 = 56	8 + 8 + 8 + 8 + 8 + 8 + 8 = 56
8 x 8 = 64	8 + 8 + 8 + 8 + 8 + 8 + 8 + 8 = 64
8 x 9 = 72	8 + 8 + 8 + 8 + 8 + 8 + 8 + 8 + 8 = 72
8 x 10 = 80	8 + 8 + 8 + 8 + 8 + 8 + 8 + 8 + 8 + 8 = 80

9	x	1 =	9

9

9	x	2 =	18

9 + 9 = 18

9	x	3 =	27

9 + 9 + 9 = 27

9	x	4 =	36

9 + 9 + 9 + 9 = 36

9	x	5 =	45

9 + 9 + 9 + 9 + 9 = 45

9	x	6 =	54

9 + 9 + 9 + 9 + 9 + 9 = 54

9	x	7 =	63

9 + 9 + 9 + 9 + 9 + 9 + 9 = 63

9	x	8 =	72

9 + 9 + 9 + 9 + 9 + 9 + 9 + 9 = 72

9	x	9 =	81

9 + 9 + 9 + 9 + 9 + 9 + 9 + 9 + 9 = 81

9	x	10 =	90

9 + 9 + 9 + 9 + 9 + 9 + 9 + 9 + 9 + 9 = 90

MULTIPLICATION TABLE OF 10

10	x	1	=	10	10
10	x	2	=	20	10 + 10 = 20
10	x	3	=	30	10 + 10 + 10 = 30
10	x	4	=	40	10 + 10 + 10 + 10 = 40
10	x	5	=	50	10 + 10 + 10 + 10 + 10 = 50
10	x	6	=	60	10 + 10 + 10 + 10 + 10 + 10 = 60
10	x	7	=	70	10 + 10 + 10 + 10 + 10 + 10 + 10 = 70
10	x	8	=	80	10 + 10 + 10 + 10 + 10 + 10 + 10 + 10 = 80
10	x	9	=	90	10 + 10 + 10 + 10 + 10 + 10 + 10 + 10 + 10 = 90
10	x	10	=	100	10 + 10 + 10 + 10 + 10 + 10 + 10 + 10 + 10 + 10 = 100

MULTIPLICATION TABLE OF
11   12

11 x 1 = 11		12 x 1 = 12
11 x 2 = 22		12 x 2 = 24
11 x 3 = 33		12 x 3 = 36
11 x 4 = 44		12 x 4 = 48
11 x 5 = 55		12 x 5 = 60
11 x 6 = 66		12 x 6 = 72
11 x 7 = 77		12 x 7 = 84
11 x 8 = 88		12 x 8 = 96
11 x 9 = 99		12 x 9 = 108
11 x 10 = 110		12 x 10 = 120

MULTIPLICATION TABLE OF
13   14

13 x 1 = 13		14 x 1 = 14
13 x 2 = 26		14 x 2 = 28
13 x 3 = 39		14 x 3 = 42
13 x 4 = 52		14 x 4 = 56
13 x 5 = 65		14 x 5 = 70
13 x 6 = 78		14 x 6 = 84
13 x 7 = 91		14 x 7 = 98
13 x 8 = 104		14 x 8 = 112
13 x 9 = 117		14 x 9 = 126
13 x 10 = 130		14 x 10 = 140

MULTIPLICATION TABLE OF

15

15	x	1	=	15
15	x	2	=	30
15	x	3	=	45
15	x	4	=	60
15	x	5	=	75
15	x	6	=	90
15	x	7	=	105
15	x	8	=	120
15	x	9	=	135
15	x	10	=	150

16

16	x	1	=	16
16	x	2	=	32
16	x	3	=	48
16	x	4	=	64
16	x	5	=	80
16	x	6	=	96
16	x	7	=	112
16	x	8	=	128
16	x	9	=	144
16	x	10	=	160

17 x 1 = 17	18 x 1 = 18
17 x 2 = 34	18 x 2 = 36
17 x 3 = 51	18 x 3 = 54
17 x 4 = 68	18 x 4 = 72
17 x 5 = 85	18 x 5 = 90
17 x 6 = 102	18 x 6 = 108
17 x 7 = 119	18 x 7 = 126
17 x 8 = 136	18 x 8 = 144
17 x 9 = 153	18 x 9 = 162
17 x 10 = 170	18 x 10 = 180

MULTIPLICATION TABLE OF
19   20

19	x	1	=	19		
19	x	2	=	38		
19	x	3	=	57		
19	x	4	=	76		
19	x	5	=	95		
19	x	6	=	114		
19	x	7	=	133		
19	x	8	=	152		
19	x	9	=	171		
19	x	10	=	190		

20	x	1	=	20
20	x	2	=	40
20	x	3	=	60
20	x	4	=	80
20	x	5	=	100
20	x	6	=	120
20	x	7	=	140
20	x	8	=	160
20	x	9	=	180
20	x	10	=	200

MULTIPLICATION TABLE OF
21   22

21	x	1	=	21
21	x	2	=	42
21	x	3	=	63
21	x	4	=	84
21	x	5	=	105
21	x	6	=	126
21	x	7	=	147
21	x	8	=	168
21	x	9	=	189
21	x	10	=	210

22	x	1	=	22
22	x	2	=	44
22	x	3	=	66
22	x	4	=	88
22	x	5	=	110
22	x	6	=	132
22	x	7	=	154
22	x	8	=	176
22	x	9	=	198
22	x	10	=	220

MULTIPLICATION TABLE OF

23

23 x 1 = 23

23 x 2 = 46

23 x 3 = 69

23 x 4 = 92

23 x 5 = 115

23 x 6 = 138

23 x 7 = 161

23 x 8 = 184

23 x 9 = 207

23 x 10 = 230

24

24 x 1 = 24

24 x 2 = 48

24 x 3 = 72

24 x 4 = 96

24 x 5 = 120

24 x 6 = 144

24 x 7 = 168

24 x 8 = 192

24 x 9 = 216

24 x 10 = 240

MULTIPLICATION TABLE OF

25

25 x 1 = 25

25 x 2 = 50

25 x 3 = 75

25 x 4 = 100

25 x 5 = 125

25 x 6 = 150

25 x 7 = 175

25 x 8 = 200

25 x 9 = 225

25 x 10 = 250

26

26 x 1 = 26

26 x 2 = 52

26 x 3 = 78

26 x 4 = 104

26 x 5 = 130

26 x 6 = 156

26 x 7 = 182

26 x 8 = 208

26 x 9 = 234

26 x 10 = 260

MULTIPLICATION TABLE OF
27   28

27 x 1 = 27		28 x 1 = 28
27 x 2 = 54		28 x 2 = 56
27 x 3 = 81		28 x 3 = 84
27 x 4 = 108		28 x 4 = 112
27 x 5 = 135		28 x 5 = 140
27 x 6 = 162		28 x 6 = 168
27 x 7 = 189		28 x 7 = 196
27 x 8 = 216		28 x 8 = 224
27 x 9 = 243		28 x 9 = 252
27 x 10 = 270		28 x 10 = 280

MULTIPLICATION TABLE OF

29

29 x 1 = 29

29 x 2 = 58

29 x 3 = 87

29 x 4 = 116

29 x 5 = 145

29 x 6 = 174

29 x 7 = 203

29 x 8 = 232

29 x 9 = 261

29 x 10 = 290

30

30 x 1 = 30

30 x 2 = 60

30 x 3 = 90

30 x 4 = 120

30 x 5 = 150

30 x 6 = 180

30 x 7 = 210

30 x 8 = 240

30 x 9 = 270

30 x 10 = 300

MULTIPLICATION TABLE OF

31

31 x 1 = 31

31 x 2 = 62

31 x 3 = 93

31 x 4 = 124

31 x 5 = 155

31 x 6 = 186

31 x 7 = 217

31 x 8 = 248

31 x 9 = 279

31 x 10 = 310

32

32 x 1 = 32

32 x 2 = 64

32 x 3 = 96

32 x 4 = 128

32 x 5 = 160

32 x 6 = 192

32 x 7 = 224

32 x 8 = 256

32 x 9 = 288

32 x 10 = 320

MULTIPLICATION TABLE OF
33   34

33 x 1 = 33		34 x 1 = 34
33 x 2 = 66		34 x 2 = 68
33 x 3 = 99		34 x 3 = 102
33 x 4 = 132		34 x 4 = 136
33 x 5 = 165		34 x 5 = 170
33 x 6 = 198		34 x 6 = 204
33 x 7 = 231		34 x 7 = 238
33 x 8 = 264		34 x 8 = 272
33 x 9 = 297		34 x 9 = 306
33 x 10 = 330		34 x 10 = 340

MULTIPLICATION TABLE OF
35

35	x	1	=	35
35	x	2	=	70
35	x	3	=	105
35	x	4	=	140
35	x	5	=	175
35	x	6	=	210
35	x	7	=	245
35	x	8	=	280
35	x	9	=	315
35	x	10	=	350

36

36	x	1	=	36
36	x	2	=	72
36	x	3	=	108
36	x	4	=	144
36	x	5	=	180
36	x	6	=	216
36	x	7	=	252
36	x	8	=	288
36	x	9	=	324
36	x	10	=	360

MULTIPLICATION TABLE OF
37

37 x 1 = 37

37 x 2 = 74

37 x 3 = 111

37 x 4 = 148

37 x 5 = 185

37 x 6 = 222

37 x 7 = 259

37 x 8 = 296

37 x 9 = 333

37 x 10 = 370

38

38 x 1 = 38

38 x 2 = 76

38 x 3 = 114

38 x 4 = 152

38 x 5 = 190

38 x 6 = 228

38 x 7 = 266

38 x 8 = 304

38 x 9 = 342

38 x 10 = 380

MULTIPLICATION TABLE OF
39   40

39 x 1 = 39		40 x 1 = 40
39 x 2 = 78		40 x 2 = 80
39 x 3 = 117		40 x 3 = 120
39 x 4 = 156		40 x 4 = 160
39 x 5 = 195		40 x 5 = 200
39 x 6 = 234		40 x 6 = 240
39 x 7 = 273		40 x 7 = 280
39 x 8 = 312		40 x 8 = 320
39 x 9 = 351		40 x 9 = 360
39 x 10 = 390		40 x 10 = 400

NUMBER SYSTEM

The system of representing numbers is called number system.

Prime Numbers

Any number with only 2 factors, i.e., 1 and the number itself, is called a prime number.

The examples of prime numbers are 2, 3, 5, 7, 11 etc.

Composite Numbers

Any number with more than 2 factors is called a composite number.

The examples of composite numbers are 4, 10, 12, 14 etc.

The number 1 is neither prime nor composite.

Integers

A number that can be written without a fraction component is called an integer.

The examples of integers are 0, -7 , 3 etc.

Whole Numbers

The set of all positive integers and 0 are called whole numbers.

Natural Numbers

The set of all positive numbers like 1, 2, 3 etc.; excluding 0, are called natural numbers.

Counting Numbers

All positive integers (1, 2, 3) used for simple counting, are called counting numbers.

Natural numbers are also called counting numbers.

Even and Odd Numbers

Even Numbers

Any integer that can be divided exactly by 2 is called an even number. All even numbers have 2, 4, 6, 8 or 0 in their ones place.

Odd Numbers

Any integer that cannot be exactly divided by 2 is called an odd number. All odd numbers have 1, 3, 5, 7 or 9 in their ones place.

DIVISIBILITY RULES

Divisibility rule of 2

For a number to be divisible by 2, it must have an even number in its ones place.

For example, 22, 26, 48 etc. are all divisible by 2.

Divisibility rule of 3

For a number to be divisible by 3, the sum of the digits must be divisible by 3.

For example, the number 12, sum of its digits are 3, which is divisible by 3, therefore 12 is divisible by 3.

Divisibility rule of 4

For a number to be divisible by 4, the last 2 digits (the ones and the tenth place) of the number must be divisible by 4.

For example, the number 120, since 20 is divisible by 4, 120 is also divisible by 4.

Divisibility rule of 5

For a number to be divisible by 5, it must have either 0 or 5 in its ones place.

For example, 320 is divisible by 5 as it has 0 in ones place.

Divisibility rule of 6

For a number to be divisible by 6, it must be divisible by 2 and 3.

For example, the number 36 can be divided by 2 and 3, therefore 36 is also divisible 6.

Divisibility rule of 7

For a number to be divisible by 7, double the last digit (ones place) and subtract that number from the number formed by other digits and if the answer is 0 or a number divisible by 7, then the original number is divisible by 7.

For example, in the number 105, 5 is the number in ones place.

By doubling 5 we get 10. Number formed by other digits is 10.

So the difference is 10-10=0, therefore 105 is divisible by 7.

Divisibility rule of 8

For a number to be divisible by 8, the last 3 digits (ones, tenth and hundred) must be divisible by 8.

For example, in the number 1560, last three digits are 560.

560 is divisible by 8. Therefore the number is divisible by 8.

Divisibility rule of 9

For a number to be divisible by 9, the sum of the digits must be divisible by 9.

For example, in the number 999, sum of the digits i.e, 9+9+9=27.

27 is divisible by 9. Therefore 999 is also divisible by 9.

Divisibility rule of 10

Any number that has 0 in its ones place is divisible by 10.

For example, the numbers 20, 400, 550 are divisible by 10.